D1313530

The Computer in Design

Anthony Hyman

The Computer in Design

Anthony Hyman

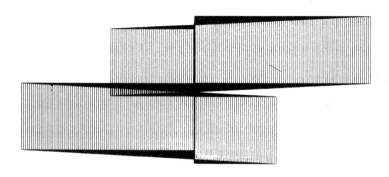

CARROLL COLLEGE LIBRARY
Waukesha, Wisconsin 53186

Studio Vista London

© Anthony Hyman 1973
Edited by John Lewis

All rights reserved. No part of this work covered by the
copyrights hereon may be reproduced or used in any form or by
any means—graphic, electronic or mechanical, including
photocopying, recording, taping or information storage and
retrieval systems—without written permission from the
publisher.
Published in London by Studio Vista,
Blue Star House, Highgate Hill, London N 19
Set in Baskerville 11 on 13pt
Printed in England
by The Whitefriars Press Ltd., London and Tonbridge
UK ISBN 0 289 70268.2

Acknowledgements

The author would like to offer his thanks to Basil de Ferranti, who lent a
helping hand as he has done to so many projects associated with computing in its brief history; and to John Carr Doughty who gave most
generously from his matchless store of knowledge of knitting.

The author is most grateful to people who have helped by providing
visual material and also in other ways. Perhaps it is appropiate to
mention: Dr Ivan E. Sutherland, himself responsible for many of the
techniques currently in use in computer graphics; Miss Andrée Rosen-
feld; Professor Peter Jewell; M. E. and R. G. Newall and T. L. Sancha
(for the computer-generated shaded pictures made at Cambridge); and
John Lewis who has designed *The Computer in Design* without the aid of a
computer.

Companies or museums that have given help with photographs are:

Calcomp Ltd (pp. 23 and 26)
Computek (p. 36)
Department of Trade and Industry, Computer Aided Design Centre (pp. 72, 78,
    80–5, 88, 90, 91 and 92)
Don Flowerdew (p. 60)
Gerber Scientific Instrument Co. (pp. 24 and 25)
IBM United Kingdom Ltd (pp. 20, 28, 29, 37, 41, 57, 58, 76 and 103)
International Computers Ltd (pp. 3, 6, 8, 19, 38, and 110)
INTEL Corporation (pp. 16 and 17)
John Carr Doughty, Design and Development Centre, Leicester (p. 98)
Mullard Ltd (p. 12)
National Aeronautics and Space Administration, Ames Research Center (p. 29)
Science Museum (Crown copyright: pp. 23, 69 and 70)
Torva Sailplanes Ltd (pp. 86–7)
Trustees of the British Museum (Crown copyright: pp. 64 and 71)
United Glass Ltd (p. 89)
Victoria and Albert Museum (Crown copyright: pp. 32, 60 and 100)
Wallace Collection (By permission of the Trustees: p. 65)
The photographs on pp. 44–51, 94–5 and 108–9 are computer-generated
objects created by Gary Scott Watkins, Computer Science, University of Utah,
during research supported by the Advanced Research Projects Agency, Depart-
ment of Defense, monitored by Rome ADC, GAFB, New York under contract
F30602–70–C–0300.

138826

Contents

CARROLL COLLEGE LIBRARY
Waukesha, Wisconsin 53186

This and other drawings scattered throughout the book were generated by a computer according to mathematical formula and defined procedure.

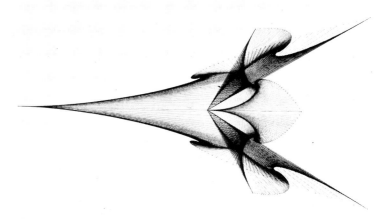

Introduction

In Renaissance Florence men such as Brunelleschi and Da Vinci were at once artist and engineer. They were deeply involved in extending both new ideas and new techniques. This unity is now a memory, a dream; the pious hope of the few. We are now in a period of unprecedented technological development. Engineers in the new industries are confronted with new materials, methods of calculation and methods of design, with barely enough time to master the lessons they have learned. In the traditional industries anxious thoughts are concentrated on a future filled with the prospect of change.

In the art schools traditional materials and methods are no longer fashionable. Indeed it is even unusual to find a sculpture course with a life class as part of the curriculum. All manner of new equipment: welding equipment, vacuum-coating equipment, oscilloscopes and power supplies confront the student. Only a few years ago there was the peace and calm of established tradition, but it is now clear that there is no turning back.

Central to discussion of the future is the computer. It is well known that computers are good at sums, both the extended calculations of mathematical physics and the tedious arithmetic of the accountant. Recently computers, having acquired effective methods of display and considerable visual processing power, are beginning to be used for design in earnest. With this development the possibility is raised of re-establishing the unity of Renaissance Florence.

But what exactly is a computer?

1. What is a computer?

A computer is a thing that does what you tell it to do. Thus it is distinguished on one side from mere tools and machines, which can be said to accept instructions only in a quite trivial sense; and on the other from the most docile of children, which can at best be cajoled into a partial and wholly transitory obedience.

A computer is a decision-making system which can alter its own decision-making program as it proceeds. The basic switching units of a computer can take many forms, being made of cog-wheels, electromagnetic relays, valves, transistors or even jets of water or air. The computer can include many different means of storing information, receiving information and despatching information. These can be dispersed over a great distance or concentrated in one place. However it is made, a computer does precisely what it is told to do: neither more, nor, unless something has broken, less. It follows that the decisive questions are: what exactly do you want the computer to do? and how precisely will you tell the computer to do it?

It is not easy to be precise, even in simple situations, and over-looking some minor point often leads to comic results, such as electricity bills sent out for £0-00. When more instinctive matters such as seeing and drawing are involved the exact statement of a problem can be very difficult indeed and require a great deal of research.

If the computer is a thing that does what you tell it to do, it is also a complex logical system which has the remarkable ability to function that way. If the main problems associated with computers today are those of using them, this was not always so. In the early days the main problem was to make them work at all.

Early methods of construction

The first computer was Charles Babbage's Analytical Engine. It was never finished. Started in the first half of the nineteenth century, and gaining the backing of the government for military calculations, it used cog-wheels for calculation like a glorified milometer or mechanical calculator. Babbage blamed the withdrawal of government funds (a contemporary note) for his failure to complete the Analytical Engine. Really it was before its time. Requiring some fifty thousand cog-wheels it did not make engineering sense.

One hundred years later the technology was ripe. First using the electromagnetic switches developed for telephone exchanges, and then the electronic valves developed largely for telecommunications, working computers were built. The first valve computers were made once again for military purposes, actually for ballistics calculations. On theoretical foundations established by Von Neumann, teams in various universities settled down to build their own computers. It was touch and go whether the first computers would work as the valves from which they were made were only just reliable enough. Lengthy calculations had to be subdivided because the computers were insufficiently reliable to complete the whole calculation before a failure occurred.

Valves are very inefficient things, with built-in heat-generators, and it was difficult to get rid of the heat. The transistor was a big advance. It is a sort of valve made inside a piece of crystal with a structure like diamond, at first germanium and later silicon. The heat-generator was gone, its function being replaced by a structure built into the crystal on manufacture. Transistors give off much less heat and are more reliable than valves. Thus more complex and reliable computers could be built.

The number of wires or connections in a computer can be huge, running into millions or tens of millions, or even more. To facilitate construction many wires are replaced by printed circuits in which alternating layers of copper and insulator, forming a three-dimensional network, replace the separate wires. It is necessary to be able rapidly and conveniently to design and make such a system, an interconnection system, without an undue number of mistakes.

A few mistakes can be corrected, but they must first be found and that can take a lot of time.

Interconnection as hierarchy: an example of computer-aided design

When making a computer it is necessary to achieve the physical realization of an abstract structure. Consider for a moment the entire logic (the abstract structure at the centre of a computer) drawn on a single flat surface. The constructional problem is to divide the logic into a family of boxes in such a way that the logic in the boxes can be realized in a hierarchy of physical structures. It is the object of advanced design to concentrate as many interconnections as possible on to the next lower and physically

Silicon integrated circuit: the vital building block. A forty-gauge cotton thread looks like a rope in comparison.

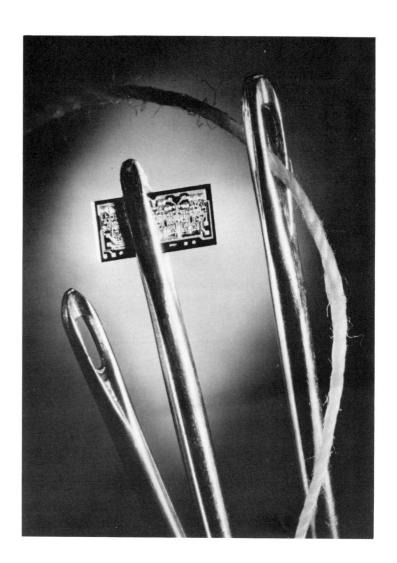

smaller member of the hierarchy, and ultimately on to the silicon slice itself.

The design of such an interconnection system is of particular interest here. It is not only an essential part of a computer but one of the most successful uses of computer-aided design and of computer graphics to date. It is also a problem close to the computer manufacturers' hearts as with such methods the computers themselves can be designed with relative ease once the techniques have been mastered. It is also the sort of problem to the solution of which computers are most easily adapted: that is to say the problem is well defined. Therefore let us consider the design of a portion of the electronics of a computer as an example of computer-aided design.

In the design of an interconnection system there are many aspects to be considered simultaneously: the reliability of each part, heat dissipation, all the manufacturing processes, the electrical rules according to which each logic element can be connected to so many others by wires, or rather tracks, of not more than a certain length, on a printed circuit board; and of great importance is the introduction of modification. It is not merely a matter of correcting mistakes, but requirements have a habit of changing at rather a late stage in design; a problem with which designers in other fields, such as architects, are familiar.

In a typical interconnection system in a computer in use today, a chip of silicon can contain an integrated circuit with two or more complete logic elements or gates. The basic switching elements are called gates because in their primitive form they opened and closed paths in different directions, on instruction. The silicon chip is directly soldered or otherwise bonded to a piece of ceramic, or plug-in. This piece of ceramic carries a number of integrated circuits, together with the associated wiring, and plugs into a connector or complex socket which is itself soldered into a large printed circuit board, or platter as it is called. The connector carries the connections joining the plug-in to the rest of the logic of the system. The maximum number of connections that can be made to the rest of the system is often more of a limitation than the amount of logic that can be physically mounted on the ceramic plug-in.

As systems become more and more compact the problem of minimizing the number of connections at each level plays a decisive part in the organization of the system. One or two bits of logic have few connections; a complete unit of a system may have only a few connections but in between the number is much larger. Therefore a decisive change occurs when it becomes possible to put an entire complex unit with only a few external connections on a single piece of silicon.

There is another interesting point here. The individual logic elements can be standard, two or three types sufficing to make a logic system. With larger units containing about fifty gates, fifty or a hundred different types may be needed to make the logic of a medium-sized computer. In general there is only one of each design of platter in each system, a given design being duplicated only where the system, or subsystem, is duplicated. A small processor, that is to say an entire small computer, is, like an individual logic element, a standard unit. It is now possible to make small computer units on a few chips of silicon.

The platter: one level of the constructional hierarchy

The largest single pieces of the construction hierarchy are the printed circuit platters. As a problem in graphics the design of printed circuits is simplified by the fact that they are usually comprised entirely of straight lines and right angles. The problem is to make connection between given points on the platter, forming a three-dimensional network, according to given rules: for electrical reasons it can be required that the tracks shall not exceed a maximum length and that neighbouring tracks shall run parallel to each other for only a short distance. Each time a track is used up it is not available for further use. Thus finding a path for the first inter-connections is easy for the computer which is being used as a design aid. As the available spaces are used it becomes a steadily more difficult problem of finding a path through a maze. It is interesting that when most of the paths are filled the eye can be a great help in choosing routes. The main point of using a computer is that there are many rules to be obeyed and information is transferred many times from one part of the calculation to another. The computer unlike the human brain does not make careless mistakes. Of course if it is wrongly instructed, if for example the electrical rules are wrong, then the results will be wrong. If you tell the computer to behave nonsensically it will do so: the phrase commonly used for such activity is 'garbage in—garbage out'.

Such a design system can itself take three or more years to

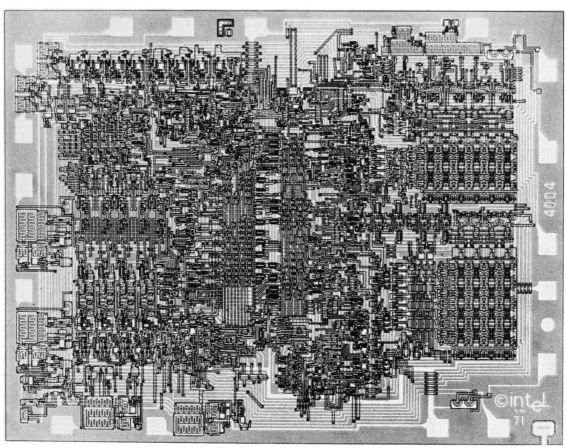

4-bit parallel processor, the centre of the computer, on one chip of silicon.
First a transistor, then a gate . . . a thousand gates . . . When made in large
numbers any silicon chip soon costs less than a dollar, no matter how much
logic it contains.

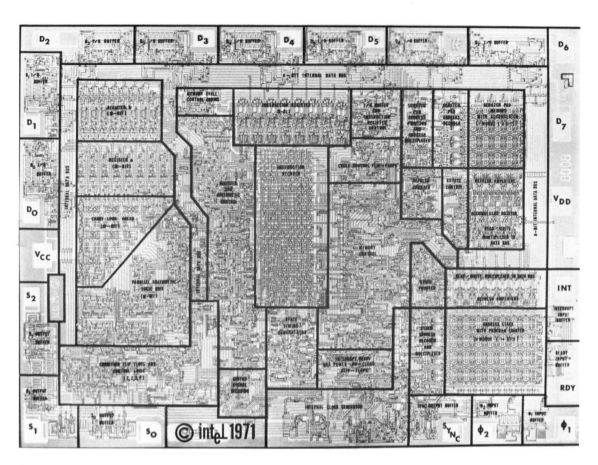

8-bit parallel processor on one chip of silicon. The functions of the computer have been drawn in. The amount of logic that can be put on a chip roughly doubles every year: exponential growth without ecological disadvantage.

develop, but should then be reliable. It is this long time scale, with the expense that it entails, that again and again makes it essential to get initial assumptions right.

Once the pattern is decided it must be realized in physical form. In computer-aided design of printed circuit platters a flat-bed plotter is used with accuracy of about a thousandth of an inch. The pen of the plotter is replaced by an accurately focused beam of light which draws patterns directly on to a photographic plate in a darkened room. The plate is then developed and fixed.

A sheet of insulating material, clad on both sides with thin copper sheet, is coated with a material sensitive to light and exposed through the photographic plate. The copper is subsequently etched leaving behind a pattern of tracks corresponding to the pattern formed by the light spot on the photographic plate. A number of such sheets is bonded together and holes are drilled by a program-controlled drilling machine where connections are required between layers. The holes are internally plated with copper.

Many different variations of this method are possible, but the result is the same: by computer-aided design an intricate three-dimensional network has been made which takes its place in the constructional hierarchy.

A series of memories

The computer also has a memory, or rather a series of memories: small memories which hold the information in immediate use (the technical terms are registers, scratch

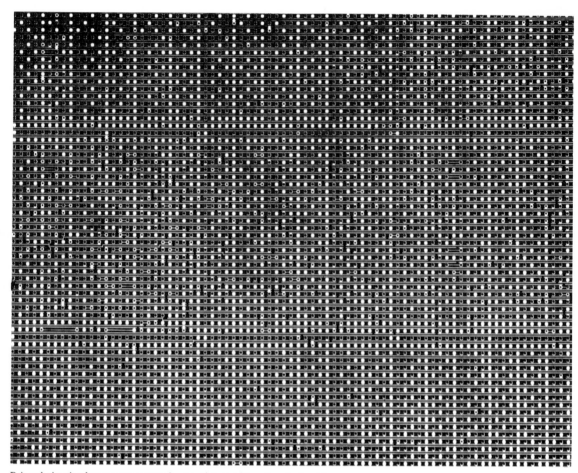

Printed circuit platter: copper tracks, ten thousandths of an inch wide, are supported on an insulator.

Designer and display: a computer is used to design part of the electronics of
a computer.

pads and so on); a rather larger and slower memory (main memory) from which instructions and facts (data) are taken and put on the small working memories as required; and much larger memories (magnetic tapes, discs) in which information can be kept, so to speak, at the back of the computer's mind.

The hardware works

A computing system may be thought of as being composed of a set of physical devices (hardware) on which are written sets of programs (software). There are many different ways of organizing such systems, but a general trend is clear: in the early days attention was directed to the fine details, but the system seems to be receding, as if seen through a zoom lens; the details of the hardware become of less and less interest because they can be taken for granted: the hardware works. Also the availability from component manufacturers, and at component prices, of things which can meaningfully be called computers permits dispersal of the system so that each decision can be put at the logical place: as close to the point of action as possible.

In general the interest is now focused on using computers in many different applications and on designing computing systems (hardware and software) so that they can be used easily and conveniently.

2 The computer draws a line

If you would draw a line you must take paper, pen and ink, or other materials. You must decide where on the paper to start your line and how it shall proceed; in what direction it shall go and of what thickness it shall be. If you would draw lines of different colours you must have several inks and several pens, or means for cleaning the nib.

Similarly the computer needs not only information as to the form of a line but means for making it. The medium is at any rate a part of the message.

Plotters

In the case of the printed circuit platters discussed in the last chapter, the lines are simple in form but the means for realizing them are specialized. An accurately focused source of light is placed on a cross-arm and is moved along the arm by a screw thread, carefully designed to avoid backlash, in steps of two tenths of a thousandth of an inch with an accuracy of one thousandth of an inch. Movement in the other direction is achieved by moving the entire arm backwards and forwards with the same resolution and accuracy. A diaphragm on the source of light opens as the head accelerates so that the exposure is independent of the rate of traverse of the head. Spots of light with a range of diameters can be provided.

The light head can be replaced by a pen which is raised and lowered on command. A pointed tool can be used to inscribe fine lines in a suitable medium. Such equipment is also used in map making.

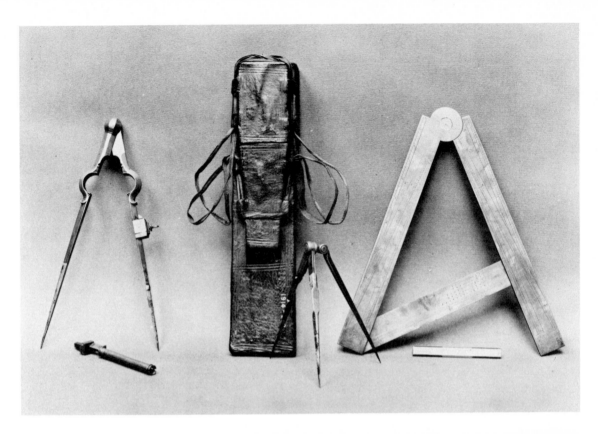

The traditonal method: early
drawing instruments (17th century).

Plotter and pen drawing a
printed circuit board.

Drawing a pattern on a film, under actual dark room conditions.

Flat-bed plotter: this one draws at 600 inches a minute over a surface of 5 x 24 feet. It can also be used in reverse as a digitizer.

Other models of plotter make larger steps so that on receiving a similar command from the computer a longer line is drawn and the line is drawn at a correspondingly higher speed. Such plotters can be twenty or thirty feet long when used for designing ships' hulls or for air frame lofting. They can also be a foot long, or even work on a microscopic scale. Thus plotters are made which work over wide ranges of speed, area and accuracy.

These are the flat-bed plotters. No one would have bothered to use such a name if there were not other types of plotter. All that is required is to move one or more points relative to a piece of paper, or other medium, and there are many ways of doing it.

In most flat-bed plotters the paper remains still while a single point moves in two directions: along the arm in one direction; and in the other direction when the whole arm moves.

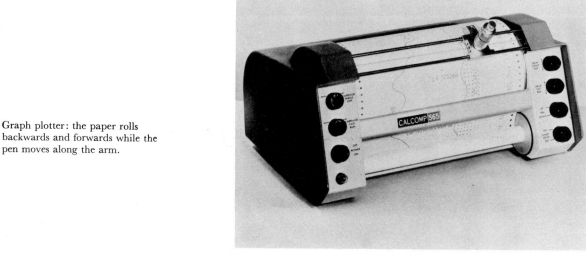

Graph plotter: the paper rolls backwards and forwards while the pen moves along the arm.

In a different type of plotter, often called a graph plotter, the pen moves along the arm, while the medium, a roll of paper, moves backwards and forwards under the arm. This has two advantages: the heavy arm which holds the pen stays still, and a long roll of paper can be used.

The next logical step is to replace the moving pen by a row of points stretching across the paper each one of which can make a mark. The desired pattern can be made right across the paper all at once while the paper steadily rolls past. Such a system can be much faster because many points can be made simultaneously, whereas a single pen can be at only one place at a time.

Many different methods of making marks on the paper have been used, In one system droplets of ink are sprayed from tiny jets. In another the ink is held on a drum magnetically and rolled on, much as in newspaper printing. A row of styli may simply hammer ink from a ribbon on to the paper. There is the Xerox system in which the ink is held electrostatically on a selenium-coated drum. In another system ink is drawn by electrostatic potential directly on to the paper.

In these systems, in which a row of points is made as the medium rolls past, a change is necessary in the form in which information is sent to the plotter. The information must be reorganized inside the computer before being sent to define the marks to be made on the paper. It is the way in which the information is organized, rather than the particular physical means for making the marks, that is the crucial distinction between types of computer plotter.

Printers

If instead of making a dot at a time one has a preformed character, as in a typewriter, then one has arrived at the conventional line printer. In the line printer a drum with letters formed on it rotates rapidly while very fast hammers hit the paper at the right time to form the required letters. As in typing several copies can be made at one time by using carbon paper. This is important for legal purposes.

In printing no particular problem arises in organizing the information as print already happens a line at a time, the letters being organized sequentially according to the conventions of our language.

Display screens

The plotters discussed above make permanent drawings and script. But presentation of continually changing information is often required, and is in practice the key to much of the power of computing systems. One looks at changing information passively when following the behaviour of an industrial plant in action. In clinical observation of a heart a theoretical model of a heart is held in the computer. This

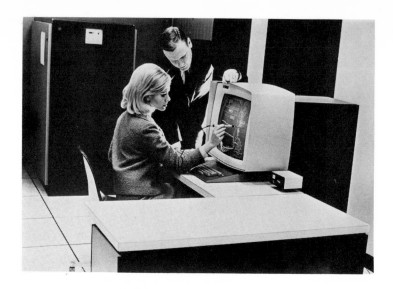

Designer and display screen: using a light pen.

model is used to distinguish and separate the picture of the beating heart (of which two views are obtained by X-ray examination) from unwanted background. In this case the viewer can alter the picture to the extent of varying the point of view from which he looks at the heart, which can be seen in sufficient detail to identify malfunction. If a computer is being used for design the information will change as design proceeds. Therefore in addition to the display screen the system must provide means by which the information (picture) on the screen can be modified and altered at will.

Almost everyone in developed countries has a display television screen in their home, possibly showing colour. At present it can only be looked at passively, control being limited to choice between different sequences of pictures made by someone else, as well as modification of the brightness of the picture and of the volume of the accompanying sound.

With the picture telephone a greater degree of interaction occurs between viewer and viewed. Picture telephones will one day be used to obtain access to public information services, the computerized version of the public library. In

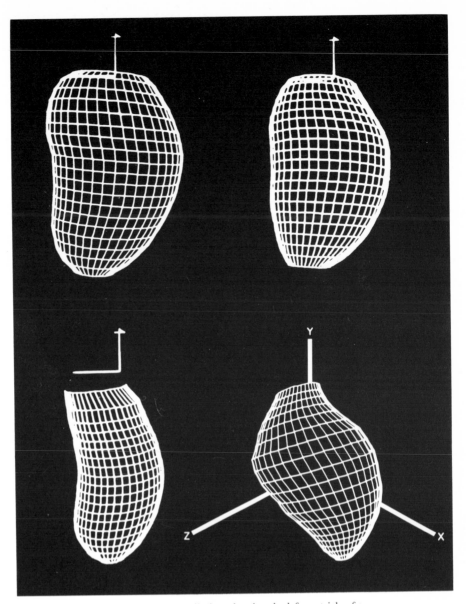

Excerpts from an animated computer display, showing the left ventricle of a
patient's heart in various stages of contraction. On the bottom right is the
expanded left ventricle of another heart which has a plastic valve.

due course they will no doubt be developed into computer terminals with other facilities. At that point computing facilities really become available for mass consumption, if that is the right word. To a generation brought up to play with computer cars, with their mothers using program-controlled sewing and washing machines, it will seem the most natural thing in the world.

The idea that computing is necessarily difficult and to be reserved for senior grades or forms at school is surely wrong, corresponding to a rather early stage in the development of computers: experimental studies with seven-year-old children have shown that children will take to the use of display terminals with ease if they are given the opportunity young enough. Today computers are still expensive and computing languages by and large still rather clumsy. (It is curious that after working with computers for many years some people have deluded themselves into thinking that ForTran, the most commonly used scientific computing language, resembles a natural language.)

As far as arrangement of information is concerned, the ordinary television screen corresponds to the making of drawings, a row of dots at a time, as the medium rolls by. Television is of course based on a scanning system in which the electron beam scans the picture a line at a time: 525 or 625 lines in the home nowadays; rather more in high resolution systems used in industry.

It is also possible with a display tube to draw lines directly as with the flat-bed plotter. This is usual in the accurate, high-resolution displays used currently in computer-aided design. Such displays can show detail which would require four thousand lines in the sort of scanning system used in television sets. Some of these displays can draw lines in several colours.

Static pictures

It is sometimes convenient to have the picture stay there once it is put on the screen. In an ordinary display tube a beam of electrons moves across a screen coated with a phosphor which glows where the electrons hit forming a light line on a dark background. Physical arrangements are also possible, as in the so-called storage tube, in which the charge pattern is stored on a fine mesh. After that the electron beam only passes through where a charge has been stored thus forming a static picture. Means for deliberate erasure are provided but, while the picture remains, the flicker, so irritating in many home television sets, is completely removed. The storage tube provides a cheap way of storing a picture which is particularly valuable at the far end of a telephone line.

Another technical means for making a storage tube uses a photochromic chemical in place of a phosphor. The tube is illuminated with light from the back. The light is not transmitted where the electron beam has hit the screen. This system draws a dark line on a light background as is customary with pen and ink.

The storage tubes just described store the picture on their faces, where it can be seen and are therefore called direct view storage tubes. Such a storage mechanism is valuable in

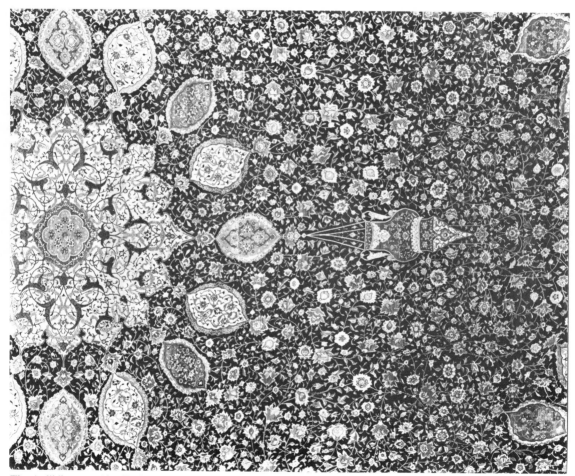

The Ardibil carpet (detail). Such design involved a simultaneous view of
detail and of the whole, requiring a resolution far beyond anything computer
display screens can handle at present.

itself and there is a neat system which combines the advantages of the direct viewing storage system (a simple form of information, cheap storage) with those of a television type display (cheapness because of mass production, shaded pictures): that is the 'scan conversion tube'. It is a special type of electron beam tube with a target, made of silicon by techniques similar to those used for making integrated circuits, which can store a charge and then, in a different mode of operation, read the stored charge without noticeably altering it. Means for erasing either a part or the whole picture are provided. The tube is scanned in such a way that the information can be transferred directly to an ordinary television screen. If three storage tubes are used, one for each colour, a colour television screen can be used. Thus there is the basis for a cheap display that will handle both shaded pictures and colour.

Choose your medium

The computer now has the physical ability to make lines in more or less any form and on any material one chooses. A wide range of physical mechanisms is available and readily controllable, from the simple pencil to an electron beam or beams of ionized molecules for etching sheets of metal directly under the control of a computer. Inside or outside the appropriate gleaming cabinets lines can be drawn, formed as a series of dots, engraved on to metal, drawn on to a photographic plate, scribed, printed on to cloth. . . . Which methods are actually developed for standard use is a matter of economics; but if you dig around in the world's laboratories almost every method which has been thought of is being used somewhere. Moreover if computer technology should become available in the schools of art and design many more imaginative ways of using these methods will doubtless be developed.

Where a real limiting factor does occur is when one wishes to display a large amount of information in a short time. It is common knowledge that the amount of detail which can be shown on the ordinary television screen is limited, essentially because the picture is changed twenty-five or thirty times a second so that movement can be shown.

What one would like is a display system in which the resolution (degree of maximum detail which can be displayed) and repetition rate (rate at which the material is repeated) are variable at will, from a rapidly changing low resolution to a slowly changing high resolution picture. Such systems are not yet available. No doubt they will be in due course: technology has more or less reached the point at which if you can specify a circumscribed system then, if it is at all plausible, someone will find a way of making it.

It is worth remarking that users of displays regularly demand higher and higher resolution. There is much to be said for designing the forms of presentation to require as low a resolution as possible. Apart from the problems of high resolution there is a limit to the rate at which the mind can digest information, and it is a good discipline to be forced to remove irrelevancies.

Defining lines to the computer

If it is simple for a computer to draw a line, the inverse

process of telling the computer where to draw a line may not be at all simple. The computer can draw all manner of types of line once it knows where to draw them, but how is it to know where to draw them? Or rather how are you to tell the computer where to draw them?

As soon as lines are well defined the problem is solved. The difficulty arises with hand-drawn lines: the essential features of hand-drawn lines have not been properly analysed and are not well understood.

It is always possible for a line to be subdivided into a very large number of dots giving both direction and thickness. That is how photography works, the dots being tiny particles of silver which form in the emulsion. But to be of much use in computing, more compact and powerful methods of describing lines are necessary; and that implies selection and choice. First of all compression of information is necessary to keep the amount of information held in the computer small; relatively that is. Secondly it is necessary to keep the number of variables under the designer's control sufficiently small for him to handle conveniently. Thirdly, and more subtly, a good case can be made that it is necessary to impose a unity on the design; but more of that later.

Compactness of information then is a great help in describing a line; when one comes to describe general curved surfaces in three dimensions a compact form of description is mandatory: one simply cannot begin without a compact form of description.

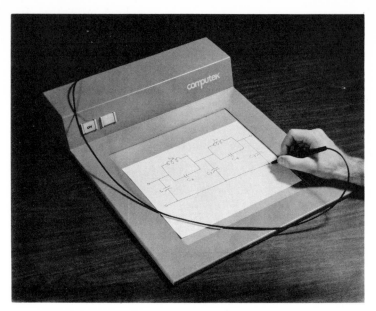

Digitizing pad: the computer takes note of a hand-drawn line.

Technical means for making the computer aware of hand-drawn lines have been developed. They are of two main classes: the light pen, and the digitizing pad. The first is really a photocell which is pointed at the television screen and sees things that are already on the screen. It is called a pen because it looks vaguely like one, and because it appears to draw with light on the screen, forming a line by dragging a cross, or other convenient mark already shown on the screen, across the screen and leaving a trail behind. The light pen is most useful for picking out things already displayed on the screen and therefore known to the computer. It is a useful tool for assembling simple elements into more complex structures.

The digitizing pad is a device on which one draws freehand while the computer notes where the pencil or other indicator is passing. Instead of looking at the pad one may look

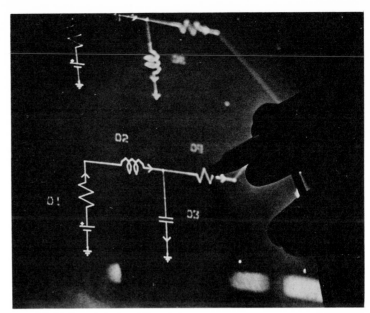

Designing electronics with a light pen.

directly at the television screen, watching a point of light, or cross, or other convenient marker which moves in sympathy with the hand. This is really what one does with a pencil, which is merely a stick linking hand and point. With digitizing pad and screen the stick is replaced by a logical link through the computer. This permits the power of the computer to be used as one goes along. The size of the picture can be varied at will, as with a zoom lens, and the picture can be built up in exactly the form in which it will actually be used, representing the lines in a chosen mathematical form.

Other devices, essentially similar to a pad but more simple and less convenient for freehand drawing, such as a rolling ball, or joystick, or simply buttons which cause the point of light or marker to move on the screen, have been used. Also pads with very high accuracy have been used for map making and similar work. With the simple pad, like the one in the illustration, there is no noticeable delay between drawing the line in a perfectly normal way and the computer's response, so the designer can interact with the computing system in a natural way, developing and building up his ideas and making use of the power of the computer as he goes along.

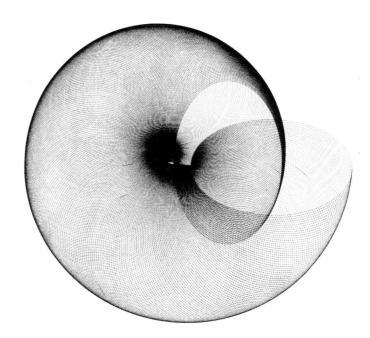

3 The third dimension

Painters have for a long time studied the problem of representing three-dimensional objects, animals, landscapes, tools, or whatever was the subject of interest, in three dimensions. The problem arose when the first cave artist drew an animal. During this time a great deal has been discovered about the possible means for achieving verisimilitude with different media and for different purposes. There is really no single solution to the problem: any method implies selection and must be suited to a particular situation, problem or set of problems. In practice a whole range of stylized representations has at different times and places become conventional.

When some convention has become sufficiently familiar it is often called realism, and the artificial nature of the form forgotten. Thus an ordinary black and white photograph of a familiar scene, once so popular with the most philistine devotees of realism, can be meaningless to a native of New Guinea unfamiliar with the medium; and by contrast the representational aspects of tribal masks are often underestimated in the sophisticated west.

Displays

In computer-aided design the problem of representing three-dimensional objects on a flat surface has to be faced in a rather exacting form. For one thing the display screens available at present, and likely to be available in the near future, can only show at one time much less detail than conventional drawings, let alone a photograph. Plotters of several types can draw with as much detail as one wishes, but many uses of computers require continuous interaction with the system with no noticeable delay, and continuous interaction means using a display.

Thus all the problems which occur in embryonic form when work is restricted to flat surfaces, to two dimensions, arise in earnest in computing when one comes to handle general three-dimensional forms.

The screens used are the same in two- and three-dimensional work, although for three-dimensional representation, two screens, or a single screen on which two series of images alternate in rapid succession, can be used and presented one to each eye so as to give a stereoscopic image.

What the computer screen can do, besides using the conventional orthographic, isometric, perspective and other drawing systems* in displaying three-dimensional objects, is, by changing the picture, to give the appearance of movement, either of the object or of the viewer. Which of the two is intended can be indicated by the provision of the appropriate visual cues.

The sort of thing done is to move around an object, or rather to present a moving picture on the screen as if one were moving around the object in a space ship, looking at the object from different angles at will and moving in and out to see different degrees of detail. One can also show a building at different stages of construction (motion in the time dimension), recalling successive stages from the computer's memory; or emphasizing different aspects such as

* See *Drawing Systems*, Dubery and Willats, Studio Vista 1972.

Moving helicopter on a display screen.

the heating and ventilating system. One might look at a whole aircraft with its aerodynamic form and then track in until looking at the thread of a particular screw. The computer store can hold information in far more detail than a piece of paper, and the current lack of detail which can be shown at once on a screen must be compensated by the ability of the computer to focus attention on each aspect as required.

Physical models

Displays have their limitations and for some purposes it is necessary to cut full-size models. Certainly the managing director of an automobile company would be reluctant to authorize production of a new model from a picture on a screen, whatever the detail shown. The computer is well suited to producing such models. Provided a sufficiently complete mathematical representation of the car exists inside the computer, and the appropriate output system is joined to the computer, an exact model can be made with great rapidity. In the *Système Uniserf* of Bézier at Renault a three-axis numerically controlled milling machine with a point cutter (he calls this a three-dimensional drawing machine) is attached to the computer. This actually cuts parts in polyurethane foam. Metal parts can then be sand-cast, using the foam as pattern. Another machine has been developed for cutting the traditional full-size models of automobiles directly in clay.

Defining general surfaces

The problem of defining to the computer the shapes required, in a manner convenient to the designer, is not simple. Three-dimensional analogues of the digitizing pad have been made, such as a wand which the designer can move while the computer notes the position of an ultrasonic bleeper on its tip. What is required is a system which conveniently combines a hand input with the power of the computer to rotate, move and change the scale of objects on the screen. The author has devised such a system, and no doubt others have, but no such system is commercially available. Nor indeed would a single system meet all requirements. One does feel that it is in solving problems such as this, developing tools for the designer, that the schools of art and design can play a big part if they are given the opportunity.

Some kinds of solid, such as cylinders, spheres, surfaces of revolution and singly curved surfaces can be represented quite well on a flat surface. The design of doubly curved surfaces has presented difficulties to the designer ever since machine manufacture became the dominant method of production. For one thing it is actually difficult to make an object with a doubly curved surface. There is also a difficulty in communication between different stages of the design and manufacturing processes, and one way in which computers can help in three-dimensional work is by defining forms not easily defined without ambiguity in any other way. Some day the widespread use of computer-aided design, together with the use of non-mechanical forming and shaping techniques, may lead to a general relaxation of this constraint.

Even now it is not always possible, for functional or aesthetic reasons, to restrict the curved portions of surface design to the above more simple forms. To make it possible to use

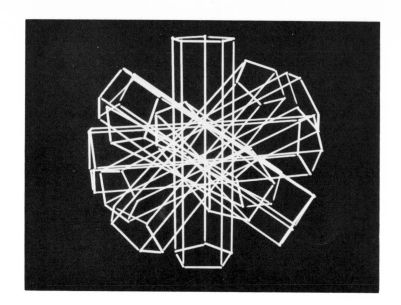

On this and the following pages are
a series of computer-generated
objects displayed in the form of both
line drawings and shaded pictures.

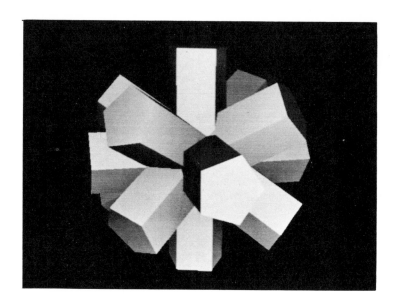

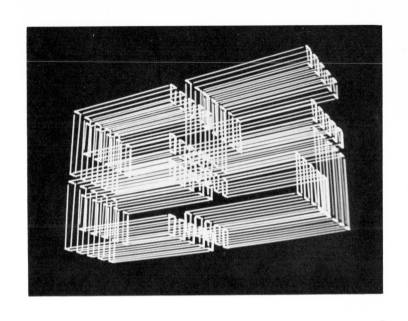

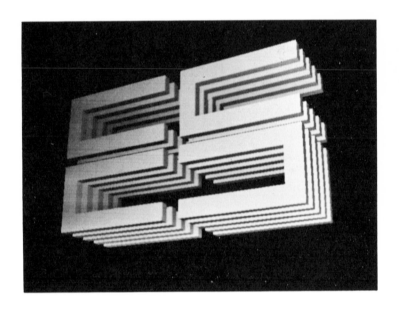

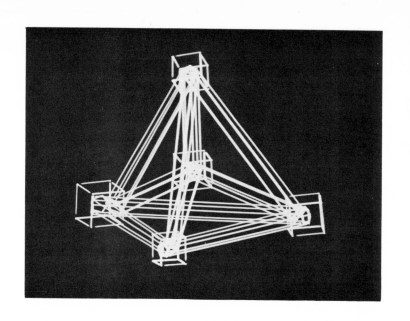

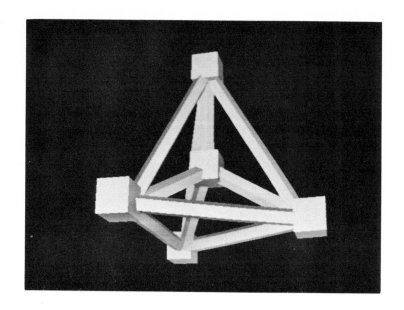

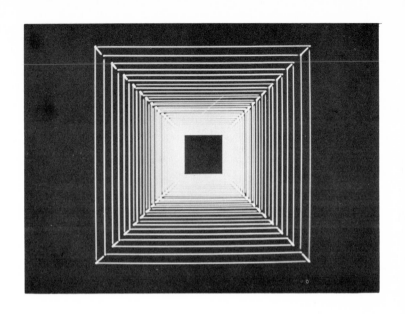

47

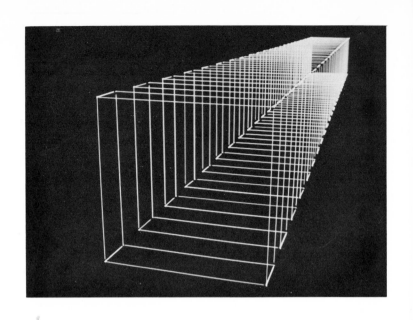

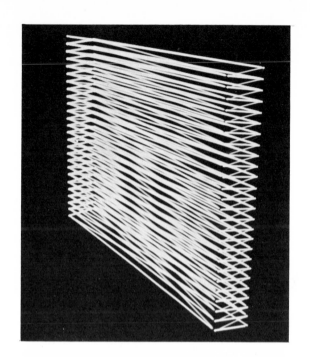

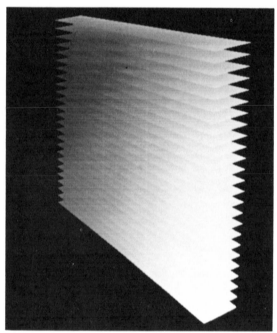

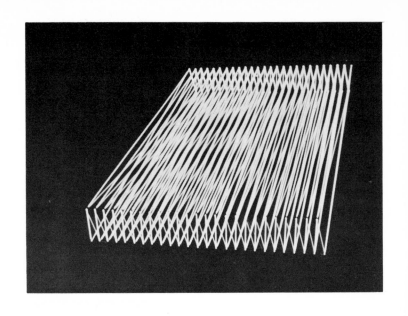

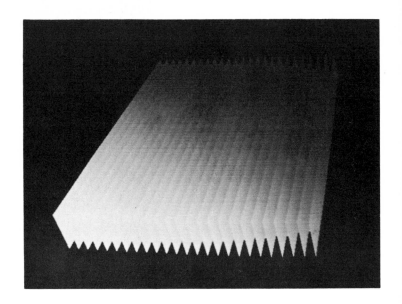

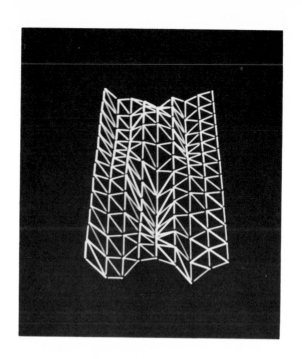

of Maintenance Processor. In this system the files are arranged so that all the data associated with each part are kept together in one place.

It is also possible to design filing systems in a somewhat different way, so that, for example, the geometrical data describing all the parts is held in one place, the manufacturing data in another, the cost in a third, and so forth. Such files are called 'inverted files' the information being sliced up, indexed and stored in a somewhat different way. A range of intermediate forms is possible.

The development of efficient filing systems has proved to be difficult and expensive, and much effort is being made to develop general systems which, at least in a set of variations, will serve a wide range of applications. These are the data base systems.

Computer filing systems may be regarded as a natural extension of methods in use in the kitchen: not in every kitchen, kitchens are known in which there is no system at all, but in a well organized kitchen. In a large canteen or in a well stocked cellar with a much larger number of items a system is essential. In a small computer the few thousand items of information must be classified carefully. With powerful modern files containing perhaps a thousand million items, any one of which may be required immediately, the most careful attention to the structure of the filing system is necessary if any item of information is to be found, and if necessary modified or erased, quickly and efficiently.

Once a reasonably efficient filing system is established it can be very difficult to improve by tinkering. On the other hand it is not difficult to make a system inefficient by bad design.

Always present is the problem that the information in a computer tends to be volatile: that is to say may easily evaporate and be lost. If the label to a piece of information in a computer is lost the information may effectively be lost: there may simply be no way of finding out where it is or what it means. By contrast the trained eye can locate a bottle and a sensitive palate define its contents.

This serves to bring out the two main methods of classifying information:
1. by directory or list, which is to say explicit address;
2. by content or attribute.
The two methods can be mixed in many different ways.

Thus: go to the third shelf from the bottom and take the fourth jar from the left (address, address); the third shelf from the bottom and take the jar with raisins in (address, attribute); the blue shelf and fourth jar (attribute, address) or the blue shelf and jar of raisins (attribute, attribute).

Depending on how it is used, a particular designation can serve as either an attribute or an explicit address. Lord George-Brown is undoubtedly an attribute, doubtfully an address. Once upon a time if you asked to see the Lord of the Manor you might, if admitted, reach the right person. The geometrical description of an object is often held in a separate area of the store, sometimes in the form of the homogeneous co-ordinates of projective geometry, the transformations appropriate to a particular point of view being defined by matrices.

It is possible to carry out such transformations locally in the file store, together with the subsequent removal of areas which would be hidden from view, and to carry out other operations, on instruction from the central processor. For example a particular part or assembly, such as the wing of an aeroplane, can be isolated from other parts for separate inspection. The resulting picture can be stored on disc or a loop of tape as in a television still and presented directly to the display screen. Such a system may be called a 'geometrically addressed file store'.

The display file, which directly controls the display screen and is quite different in form from the conventional coded display file of line drawing, can require a large amount of information. Thus it makes sense to provide alternative data paths directly from file to screen without passing through the central processor. Each change in the basic data will be immediately reflected in the display file and thus on the screen.

If in turn the user of the screen is to be able to interact with the system the display file must include pointers back to the data structure from which it was derived.

At present the physical structure corresponding to the most logically powerful file store is the rotating magnetic disc. Such file stores are already computing systems in their own

3330 disc file.

right. As it becomes economically possible to associate a small but fast computer with each head, or group of heads, the old idea of a self-contained content-addressed file store, selecting information according to one or more attributes as it passes under each head, becomes feasible.

For example, if information is both defined and searched for by three attributes at once the information required and no other will be passed to the central processor, which merely has to define to the file store the attributes of the information required and accept it when it is found.

Bits per inch

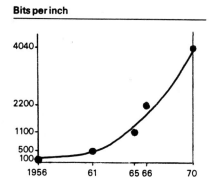

'Head Flying' Height

(Micro-inches)

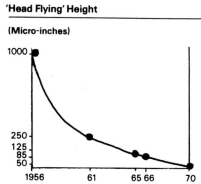

Tracks per inch

Graphs showing the advancing technology of disc files.

With one head per track, each head having its own miniature computer, file stores become very powerful indeed, and such specialized systems as the geometrically addressed file store can be realized.

Processing is being placed locally, where it should be, in this case in the file store itself instead of inefficiently sending large amounts of information through the central processor.

Much thought is being devoted to improving both the logical and physical structure of filing systems, but there will always be a considerable advantage in keeping the files themselves as small as possible. Again generality and efficiency pull in opposite directions. In particular, decisions must be taken about the design forms to be handled. There will be almost as many views of the choice of design forms as there will be designers: computer-aided design does not imply a particular aesthetic. What is inescapable is the need for choice: deliberate choice at as early a stage in developing a system as possible.

Symmetry: harmony. Torso, bronze poli by Anthony Hyman.

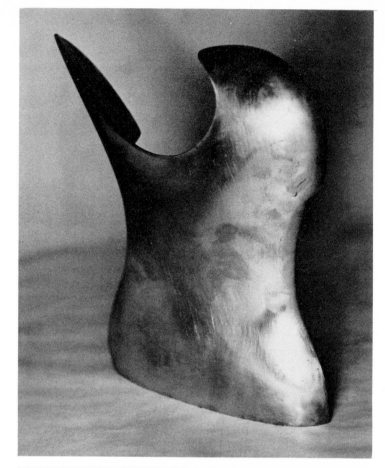

Symmetry: formal symmetry. Reticella lace (detail). Italian 16th century.

4 Design: technical and aesthetic

'If anyone thinks it important to a civilization that a common ground between art and science shall be found, then he had better look for it in front of his nose; for it is ten to one that he will see there something which has been designed.' Thus David Pye in his book *The Nature of Design.**

In computer-aided design these relationships, between the aesthetic and technical side, must be made explicit, must be clearly and precisely formulated.

There are two aspects of a design to be considered: the visual and physical aspects of the design as understood by the designer; and the mathematical representation of the design and associated information which is the form it takes inside the computer. This dichotomy is reflected in the two meanings of the word symmetry: harmony or balance; and formal mathematical symmetry. It has proved a curiously difficult matter to keep the two aspects developing coherently in a manner useful to the designer.

It is characteristic of computers that each time they intrude into a new subject they force practical decisions on questions which have hitherto been considered of a philosophical character, more suited to meandering discussions over cups of coffee far into the night than to the prosaic daylight world of industrial practice.

For example: should a design ideally be formed in the mind of one person or several people? In developing a computer-

* Published in this series by Studio Vista 1964.

aided design system a decision must be taken, as on the answer will depend the way in which control is distributed in the system. Once again the earlier such decisions are taken the better.

Fairing

A good example of the reasons why the aesthetic and technical elements in design must be considered together is implicit in the term 'fair', used to describe fluid dynamic surfaces in products such as air-frames, aero-engines and ships' hulls. It is required that some surfaces be 'fair' in order that they should work properly, that the air or water should flow over them in the required way. But fairness can only be judged by inspection of the surfaces, by aesthetic judgement. Thus we have in its clearest form the need to keep the data structure and mathematical methods of design, and the visual or aesthetic aspects going hand in hand.

The use of 'fairing' is not a reason for ignoring precise technical methods in so far as they have been developed. One might perhaps say that the use of fairing represents the use of that part of knowledge which comes from experience and has not yet been precisely formulated. But there will always be knowledge that has not been precisely formulated. Certainly there has been in the past: indeed these problems existed in typical form long before the complexities of modern industry.

Traditional crafts

The two aspects of design are related in many different ways.

Several crafts show particularly clearly the interaction of manufacturing technique and design form. Such are basketwork, pottery, weaving and knitting, and the manufacture of flint instruments by early man. Weaving leads merely to a flat form, going directly to the final shape only in the case of very simple garments such as shawls. The other four lead to fully three-dimensional forms for different purposes. Knitting leads naturally to fashioned garments such as the fisherman's jersey; basketry to containers; the coiling technique to pots which if left as formed have a characteristic appearance; and flint flaking to tools and weapons.

In the manufacture of flint instruments several distinguishable techniques have been used: impact flaking, which is direct flaking by blows with another instrument, either a stone or in its later and more controlled form with a softer tool such as a deer's antler; punch flaking, in which a punch is hit with another stone used as a hammer; and pressure flaking which is a much more controlled process. In pressure flaking the tool is roughed out by impact flaking or, in the beautiful Egyptian predynastic flintwork of which the fish-tail knife shown here is an example, by polishing prior to flaking. Each technique leads to a quite distinct pattern in the finish.

Learned archaeologists have methodically chipped away until they had acquired a certain skill in this most traditional craft, and it is suggested that the flints from a very early date have a finish considerably beyond that needed purely for effective use. It is as if they were being given a decorative finish. Quite possibly the love and care devoted to embellish-

Hand axe made by impact flaking.

Fayum fish-tail knife made by
pressure flaking.

ment gave a degree of perfection beyond that obviously functionally necessary which ensured consistency of function. If you are using spears against dangerous animals or hunting for scarce food there will be a fairly rapid selection against the tool makers whose spears fail even occasionally. Whatever the precise reasons a decorative aspect would seem virtually from the beginning to have entered into the things man made.

It is not always realized what a long way back in prehistory controlled design goes. The hand axe shown here is contemporary with Swanscombe Man, thus preceding on the evolutionary scale *Homo sapiens sapiens*, or modern man. It is made by the simplest of the techniques, impact flaking, but the technique is well controlled and the result attractive. A case can be made that there is already an aesthetic element in the design: when it looked right it would work well; and by the time of the beautiful fayum knife worked very well indeed.

It is not necessary that design should directly reflect manufacturing technique. A pot made by coiling may be smoothed so that it appears to have been thrown; and conversely a coiled motif may be painted on a thrown pot. The design of the Sèvres turquoise porcelain basket is taken from the design resulting from a manufacturing technique using the entirely different materials of an ordinary basket.

The important point is that whatever the design forms chosen, whatever the relation to manufacturing technique, choice is necessary. As always in developing a computer-aided design system the earlier that choice is made the better.

Sèvres turquoise porcelain basket.

Representational motifs

As well as the purely decorative there has always been a representational side to design. The two are only partly separable. Representation may perhaps have less importance than it once had, but it is open to designers to use representation in both decoration and form and it cannot simply be ignored.

One way in which representation has entered design is through the gradual stylization of designs until they become almost abstract, and may come to be regarded as purely patterns which have lost their original symbolic meaning. This trend may be seen equally well in Turkish rugs and Chinese calligraphy.

'Vers toutes les traditions réunies' – Sainte-Beuve

Until recently designs usually tended to change rather slowly, and strong local traditions were the rule. The strength of local design in Persian rugs is famous, and examples could be cited more or less without end.

It is hardly possible to grasp just how slow the movement once was of people, goods and ideas. Unless an article could be transported, if you wished to see it you had to get up and go. If you would see the work of the Master of Autun, to Autun cathedral you must go. Travel was slow and often dangerous. Moreover once seen it must be remembered and the visual memory is notoriously inaccurate.

We live today under the impact of so many information media that it requires an effort to imagine what the absence of printing, both of text and graphics, meant. In André Malraux's 'museum without walls' high quality reproductions have led to the comparing and contrasting of all styles, and all sizes of objects, from all places, over long spans of time. A further stage is reached when design motifs are put on a computer and they must be defined and quantified: they can then be explicitly contrasted and compared.

The computer in art

To avoid possible misunderstanding a word should be said about the use of the computer in fine art. This may be illustrated by comparison with the part once played by the camera. Quite apart from its use in the 'museum without walls' the camera has had two main functions: it has been used as a tool, and it has freed the painter from much of the pressure to concern himself with idealized naturalism. Similarly the computer is not only a flexible tool. No less important it frees the artist from much of the need to concern himself with the purely formal and abstract sides of art. Thus he can concentrate on . . . whatever it is he should be doing: just the opposite of the usual line of thought. Such is often the case.

The division of labour

In the beginning, when people made things for their own use, design and use had an instinctive harmony. When things are made for exchange with other things designer and user are no longer the same person, and the suitability of design for use is called in question; until today learned enquiries are put in hand to discover, if possible, whether a building which has been in use for some time does in fact function properly.

Not only do designer and user become different people when goods are made in quantity for exchange, the actual processes of production are subdivided in the interests of efficiency: the division of labour. The division of labour has forced many people to do dull, seemingly meaningless jobs. It has also led to difficult problems of technical communication between the several parts of the system of production. These difficulties must be overcome in developing computer-controlled systems because design cannot be isolated from method of production.

If it is now possible to envisage the development of integrated design and production systems it is only because the individual parts have undergone a long process of separated development. The basis was laid for Babbage's Analytical Engine in the development of weaving, as well as mechanical engineering and the craft of the clockmaker. The complex chains of gear wheels came from the clockmaker. So also does the idea of fixed stores which were used to control the striking of the hours and the complex musical patterns of the carillon.

Babbage's punched cards come from the punched cards of Falcon's loom and the later system of Jacquard. The numbers in the Analytical Engine were to be woven together in the mill, in any pattern required, and kept in the storehouse until required for further use or despatch to the customer. Babbage's friend, the Countess of Lovelace, said 'The Analytical Engine weaves algebraic patterns, just as the Jacquard loom weaves flowers and leaves.'

The Scheutz Difference Engine—inspired by Babbage's Difference Engine,
the precursor to his Analytical Engine.

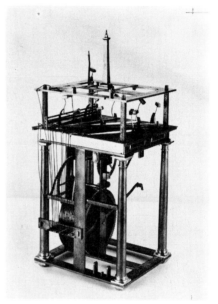

Falcon's loom, 1727 (model). Probably the first machine of any kind to be controlled by punched cards.

Nicholas Vallin's carillon clock: pin-barrel fixed store controlling the carillon.

Babbage himself saw that his Analytical Engine would permit the extension of the division of labour, previously formulated in Adam Smith's theory of pins, to abstract work. Correspondingly the stultifying effect of the division of labour can readily be observed in painfully clumsy attempts to make one 'software package' work with other 'software packages'.

Thus again: the primitive unity of design and fabrication was instinctive. Occasionally, as with the teams that built the Gothic cathedrals, living together, working together in a common idiom, with common ideas, to a common goal, a high degree of unity has been achieved in a production or construction system. Technique and design develop in harmony: out of the need to strengthen the roof and hold up the walls came fan vaulting and flying buttresses. A similar unity is found in the medieval castles.

How is such unity to be achieved in the fragmented design and production systems of today? The computer is the key. Each separate part of the design and production processes must be analysed and unity achieved by carefully stating all the problems (T. R. Thompson's dictum: 'state the problem'), by selection and by planning the system as a whole.

Who will develop such systems, crossing as they do all the artificial but heavily fortified frontiers of the professions, is still an open question.

Nicholas Vallin's carillon clock, 1598.

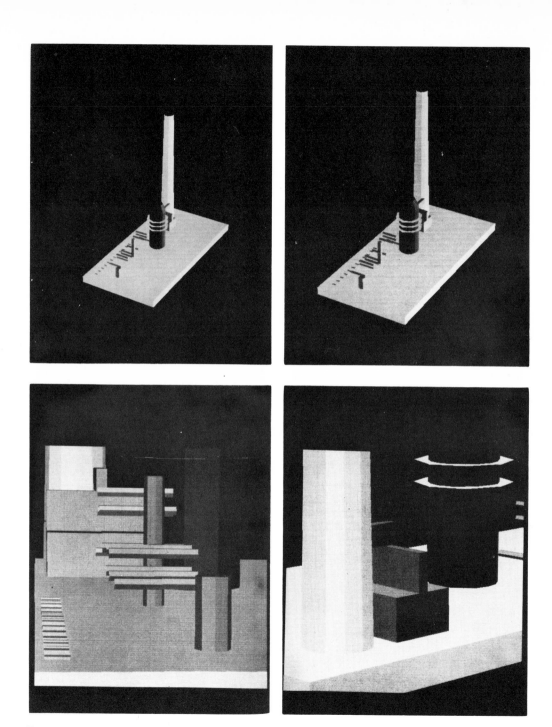

Tracking in

5 The computer aids design

Computer-generated picture of a chemical plant

Many a housewife would be surprised to see how small a radiator she really needs in the sitting room, instead of the big one now cluttering up her wall; and many an architect would be surprised to see how small a boiler he really needs to service the district heating system in the block of apartments he has designed, instead of the huge one which has just been recommended to his clients. The calculation of heat loss and the specification of the associated pipework and heating equipment is tedious, and in most cases the short cuts taken lead to gross overestimation of the size of equipment required: if the equipment is too large it will probably not be realized, but if the home is cold there will be complaints. The problems of heating and ventilating systems, the heart and lungs of a building so to speak, are well defined and the sums involved trivial for a computer, provided only that the appropriate computing resources are conveniently available.

The computer permits optimization of the use of resources, of which architecture and civil engineering are together one of the largest consumers. At a time when the resources of the planet are obviously running out the importance of this can hardly be overestimated.

The computer is neutral: like a rifle it can be used in any cause, good or bad; but at least the possibility of optimizing, in increasingly complex situations, the use of increasingly scarce resources is now there and is being actively pursued. None too soon either!

Thus the first way in which the computer is used to help in design is by doing arithmetic. There is no doubt that computing can be used extensively to do simple calculations with advantage now that cheap computing power is becoming readily available both with time-sharing and, more especially, the development of powerful miniature computers sufficiently cheap to be used in the smallest design office. Such miniature computers may be used either independently or as part of a larger system. It may further be noted that the facilities available with these miniature computers are increasing very rapidly indeed, both in scope and power and, primarily owing to the development of integrated circuits and arrays in the semiconductor industry, this trend is set for a long time to come. The limiting factor is the skill to use them.

The user of a miniature computer will have the computer, at least for a time, entirely at his disposal. In the early days of computing it was common practice for a user to sit in front of the computer which would for the time be entirely devoted to his use. With more powerful and expensive computers this practice was very inefficient because the computer might be left waiting most of the time while users got ready for the next operation. A technique was developed (multiprogramming) with which the computer could handle several programs effectively simultaneously; and a bureau manager was brought in to run the computer as efficiently as possible. A user would then send his programs and data (input) along, usually in the form of a stack of punched cards or a reel of paper or magnetic tape. In return he would receive as output other punched cards, tape or typescript (printout).

However a conflict of interests had developed between the bureau manager, who was concerned with optimizing the use of his system, and the user, who wanted results as soon as possible.

Time-sharing, a new type of operating system, was developed to deal with this problem, an operating system being the assembly of techniques by which a computer does its internal housekeeping. In a time-sharing system the computer devotes a short interval of time to each current user in succession. It should then appear to the user as if he has a whole system continually as his disposal, with great advantage.

Although some of the first time-sharing systems were inefficient, they now work quite satisfactorily, particularly when miniature computers are incorporated in the users' terminals. Indeed the advantages of immediate response which follow from time-sharing are overwhelming in such cases as the larger university computer bureaux: in such cases other systems may now be considered obsolescent.

Computing systems do not have to be all in one place. As is well known the telephone system can make remote connection. However high speed data links are expensive, and even where connection is made to a remote powerful computer time-sharing its facilities among many users, because the use of common central files or special computing facilities makes access to a central computer mandatory, the local terminal will itself tend to become a small computer. Recourse is made to the central computer as seldom as possible and the amount of data transmitted kept as small as possible.

370/165–8, the typical time-sharing computer: principal calculator in the
University of Cambridge.

All decisions should be taken as close to the point of action as possible, in this case the user at the remote terminal.

Perhaps the best known uses of computers in design have been in the aerospace industry. Certainly it is in aircraft and rocket design, and their associated equipment, that computer-aided design systems have been developed furthest, with automobile design and ship design as runners up. In aerospace developments many calculations are too complex to be carried out without computers, so that there was a compelling incentive. The practical realization of the systems was possible because the technical skill and financial resources were both available. The bulk of the mechanical engineering industry, of far greater economic importance but more conservative, has been slow to take advantage of computer-aided design.

Somewhat closer to everyday life computers have been used to do the calculations for prestressed concrete structures, such as road bridges. This work has been greatly assisted by the ability of the computer to make designs conveniently available for visual inspection. The computer has for a long time been able to draw lines, in any colour or mixture of colours. More recently it has acquired the ability to make shaded pictures.

Shaded pictures

There is a slight problem in the choice of name for these pictures. A range of grey levels is used, sixty-four levels being sufficient for most purposes. Chiaroscuro, or light and shade, is the term used in painting; but it has a slightly different

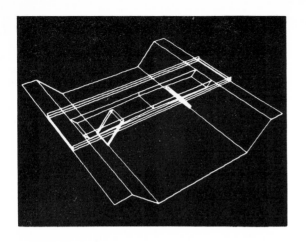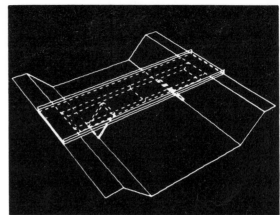

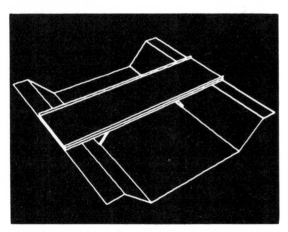

Pictures of road bridge

    Line drawing.

    Hidden lines dashed.

    Visible outlines only.

    Shaded picture.

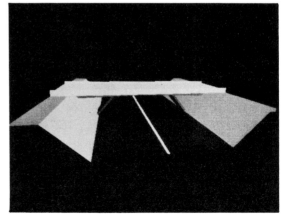

meaning which it would be a pity to blur, and in any case the term is not widely known to engineers. Incidentally that fact in itself indicates the width of gap between engineer and artist. We refer here simply to pictures drawn in light and shade, as opposed to line drawings. The greater realism of light-and-shade pictures is obvious.

Much discussion has centred on the problem of presenting drawings with sufficient rapidity, which can be a limiting technical condition. In any practical system it is likely to be necessary to limit the rate of presentation of information so that it can be comfortably assimilated. The rapidity with which a designer using a powerful computer graphics system can alter the forms themselves and his point of view may easily lead to his becoming confused and disoriented. Considerable care is required in defining stylized cues, such as a horizon or familiar object, to replace the range of cues which we use in practice to define the real world. As well as using cues a designer may find that more realistic pictures are in themselves a great help. Even so presentation of too much rapidly changing information can be quite hypnotic.

The shaded pictures show a number of simple objects presented on the front of a display tube; the objects are shown as a number of small surfaces giving an effect faintly reminiscent of cubism.

An example of design

Each design problem is really a study in itself, but it is worth looking at one example in slightly more detail. Let us consider the design of the simple bottle illustrated. The design

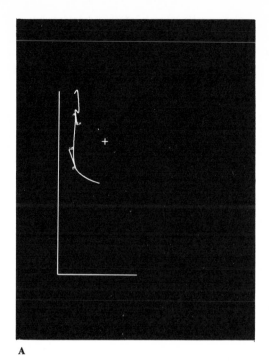

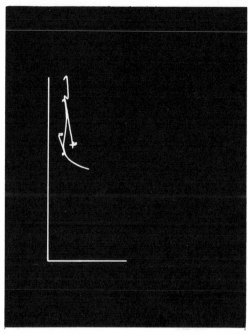

A

B

program handles a particular mathematical form of curve: to be precise a parametric cubic spline. A drawing is thus built up, when using this particular program, as a series of spline sections. The drawing is built up according to the visual choice of the designer. But other programs use other mathematical forms, and the final result is not unaffected by the choice of basic mathematical form.

The ends of the several component curves can be moved, continuity and smoothness being maintained by the system. The individual curves can themselves be further adjusted.

Figure A shows one section of curve being adjusted to Figure B, when the adjustment has been made. The two tangent lines shown are slope vectors with which in this particular system a designer controls the shape of the portions of curve.

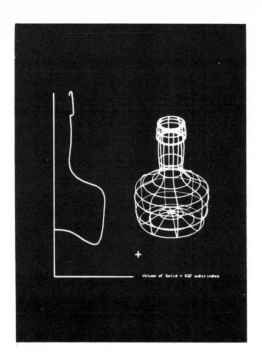

C

Figure c: the completed section is on the left. On the right is the whole bottle: that is to say the volume generated by revolving the section about the vertical axis shown. The lines are dimmed according to their distance from the viewer to indicate depth, one simple but quite effective method of making solids appear three-dimensional. The volume of the solid is automatically calculated and written at the bottom.

The next illustration shows a bucket of oil with pseudo-ripples spreading out from the centre. This object is defined initially as three separate sections: the table, the bucket and the oil. The sections are then assembled and, as they are coaxial, shifted to the correct positions along this axis. The solid of revolution is then generated. The surface reflectivity of the three materials is defined, the position of the light source indicated, and the object displayed. The position of the source of light and of the observer can be moved at will.

To display the bottle and glass on a tray the procedure is similar, but as the three objects are not coaxial the solids of revolution must be generated by rotation about their individual axes. Note the transparancy of the glass and the reflections in the glass. To form the flame the axis of revolution is bent.

The shadow of the pawn is generated by projecting a dark pawn on to the table.

Doubly curved surfaces

The objects discussed above are very simple in that they can be described as an assembly of solids of revolution. Most objects can not be described so simply and more powerful methods are required: that is to say the system needs the ability to handle general doubly curved surfaces. What is done in computational geometry is to describe the surface of a solid as a set of patches, like a patchwork quilt, each patch being defined by a simple equation, typically a parametric bicubic. This is the 'Coons Patch' which is the basis of major systems in use at General Motors, MacDonnel Douglas, Boeing and British Aircraft; as well as the straightforward interactive system called 'multipatch', developed by Andrew Armit at the University of Cambridge, which was used to design the car bodies illustrated here.

Although relatively straightforward this system is by no means ineffective, but it does require some knowledge of the theory of patches on the part of the designer. In general more complex systems are required to make the designer's task more simple.

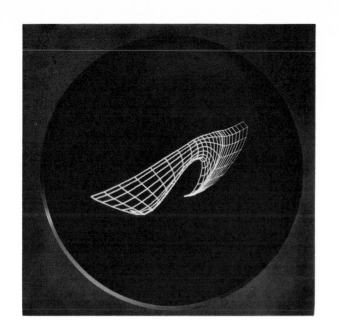

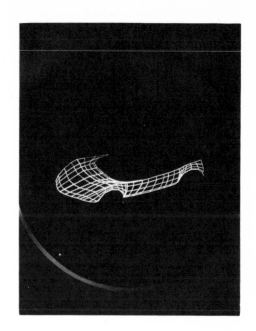

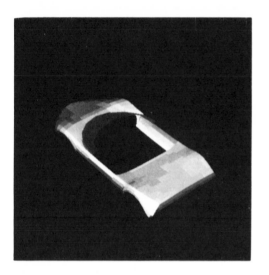

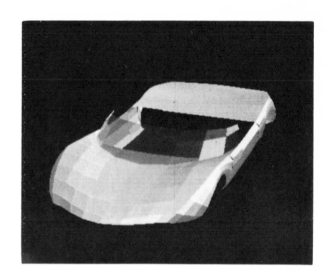

Torva sailplane.

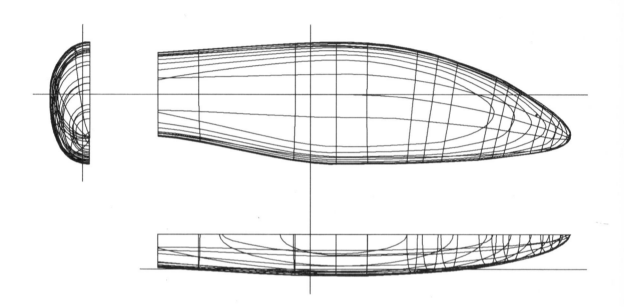

Fuselage of sailplane designed with computational geometry.

The great weight of effort has hitherto gone into developing systems that work. But it must be stressed that the visual strength of the resulting design is affected by the choice of mathematical form for the basic patch (compare the observation that the strength of a painting lies in the brush strokes).

Other mathematical forms than the parametric bicubic have been investigated. The problem does not have a unique solution as different designers' preferences and the purposes of the system will vary. In the long run the criterion must be that designs acceptable to the designer can be developed in a manner convenient to the designer.

A good case can be made that to achieve satisfactory results the ultimate user must have a say in the development of the design systems themselves. At any rate as, in a wider context, Isaac Auerbach has moderately remarked: it is an 'error . . . that the user has not taken a strong stance in what he is buying. What the manufacturer makes, the customer takes, whether it solves his problem or not.'

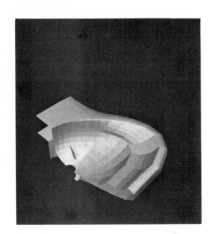

A common source of error in surface design is at the meeting of two surfaces, or the accidental fouling of two surfaces separately designed, such as the engine of a car and the bonnet. Such errors are difficult to eliminate from a formal design system and can easily pass unnoticed without a realistic form of display.

In the design of the wash basin shown in the illustration an error, obvious on visual inspection, arose when the back and front surfaces, which were separately designed, were fitted together. Similarly a sharp edge may be seen just above the dimple in a simulated design of the familiar Dimple Haig whisky bottle.

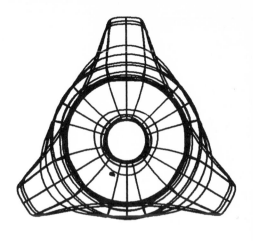

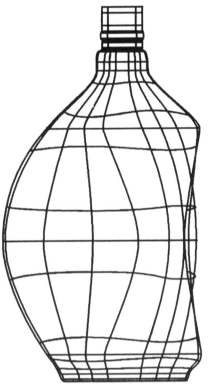

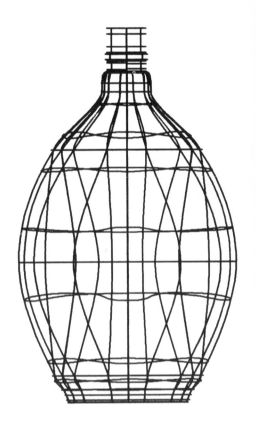

Computer-generated picture of flats . . .

. . . on demand the background appears.

Architecture and civil engineering

Computer-aided design has been used quite extensively in civil engineering, for example to calculate the behaviour of prestressed concrete structures. In civil engineering and architecture visual presentation is particularly important. Some years ago the General Electric Corporation developed a system which presents a colour television output of, for example, the driver's view as he drives through a town, showing such things as bridges looming up and passing overhead and the effect of buildings overhanging the road. It is no small increase in realism to be able to drive at will through a town before it is built and quite a different thing from having a film of a single simulated drive.

This system, which used a special-purpose piece of hardware, has stimulated many other developments. The shaded illustrations shown here are developed by software rather than hardware so that each picture is produced very much more slowly.

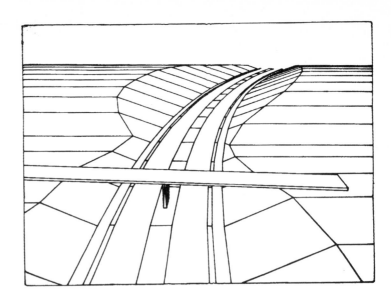

Motorway and bridge . . .

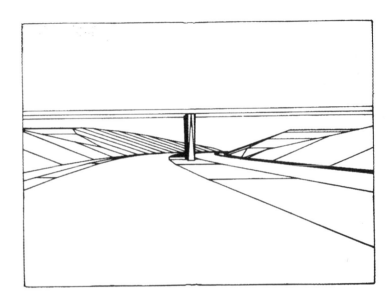

Also, a television based system could be an invaluable help
in bringing home to the general public the visual implica-
tions of new buildings and major public works, such as

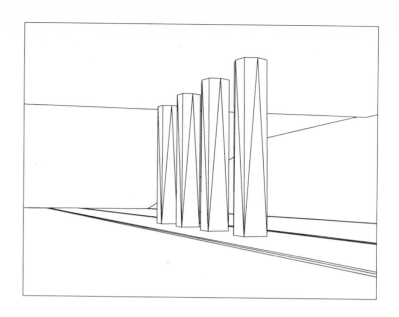

Tracking in . . .

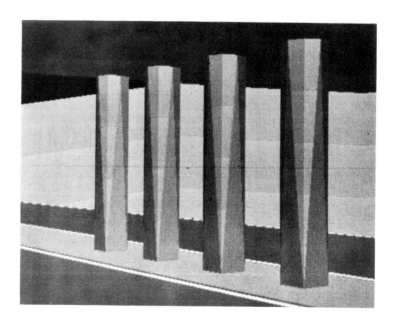

motorways, perhaps with the appropriate sound effects, before they are built.

Several integrated systems are in use for motorway design. Typically one system can be used to calculate the amount of earth to be moved when cuttings and embankments have to be made. The level of the road can then be chosen to minimize the total amount of earth to be moved. The system will draw contour maps of the area, perspective views of the road, as illustrated, sections along and across the road and so forth, on request. Some estimate can be made of the structural stability of the whole system, including the geological structure and taking into account such factors as drainage.

One essential feature of the display system developed at the University of Cambridge is that objects designed in different ways can be assembled in a standard form from which the several types of visual output, shaded pictures, line drawings with or without hidden lines, stereoscopic pairs and others, can conveniently be generated.

One of the most comprehensive systems for architectural design is ARCAID, the Architect's Computer Graphics Aid, being developed at the University of Utah. The system uses displays, and a whole range of input devices is available to the designer. ARCAID envisages the design of a building, from concept and outline scheme right through construction, without the use of paper.

Much more simple systems currently in practical use are restricted to the placement of already designed prefabricated

Computer-generated picture of office block.

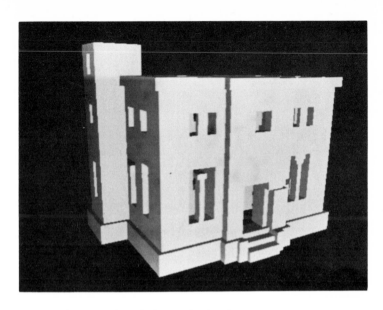

Computed-generated tonal picture of a church.

parts. A typical one is the simple system developed at the Royal College of Art for the design of schools.

It may be worth remarking, although it may well be obvious by now, that the development of the more comprehensive and powerful systems is difficult, time consuming and very expensive; so that once one or two successful systems are established they are likely to be widely adopted. This in turn will prove a strong incentive to the adoption of the range of computers on which they have been, or are likely to be, implemented.

There can by now be very few fields of design which some-one, somewhere, is not investigating with the idea of applying computing techniques. One field which is proving fruitful is the design of textiles. Textiles are economically important: everyone wears clothes and uses cloth. The technology is relatively simple. A short design cycle time is obviously of value. The technical and aesthetic aspects are inextricably mixed. Besides Babbage took ideas, as well as terminology (still in use) and his card input system from the textile industry, and it seems right that the Analytical Engine, arrayed in all the splendours of modern technology, should return to its sources of inspiration.

6 The shaping of things to come

It may well be thought that unified design and production systems belong to the far distant future. This is not so. There is no technical reason why such systems should not be developed during the next ten years: this is not such a long time scale for a major technological project and comparable with that for a new airliner. In some areas it may be said that the basic technical foundations have already been laid, but a working system would still have to be approached through a whole series of intermediate stages.

Weft knitting

How, in briefest outline, might such a system appear in the case of weft knitting? Garment making remains as the labour-intensive part of the textile industry, ripe for development. As a matter of fact advanced technology is now beginning to spread into textiles as corporations, finding less funds available for space technology, develop new fields of activity.

In the CRIF concept of John Carr Doughty (prophet and inspirer of the knitting industry) the basic idea exists for a high speed production method leading to fully fashioned garments and well suited to computer control. Nowhere would an integrated design and production system, with its short cycle time from design concept to goods in the shops, be more welcome than in the fashion-conscious rag trade.

There is a good deal in the Doughty-CRIF (contra-rotating intermittent feed) concept. Hitherto circular knitting machines, which have a high production rate, have usually formed long tubes of material and have been unable to produce fully fashioned garments. On the other hand the

The unique knit fabric designs produced by the Doughty-CRIF circular machine knitting system.

more versatile flat-bed reciprocating knitting machines are slow. Roughly speaking, the circular knitting machine is to the flat-bed knitting machine as the turbo-jet is to the piston engine. With the contra-rotating system the circular knitting machine is brought nearly into perfect balance. Using electronic needle selection it is naturally suited to computer control.

A knitting machine of the CRIF type will permit fully fashioned three-dimensional shaping, forming the shape and pattern by the same principal mechanism of needle selection. Thus pockets can be formed integrally, as well as properly selvedged edges. Any production system will lead to detailed designs special to the system: some which arise from the CRIF system are illustrated. But the potential of so versatile a system, permitting the simultaneous development of design and form, is very large and can really only be established by using it in practice.

Three-dimensional seamless knitting is an old craft and fashion-conscious small girls may be observed wearing Peruvian headgear, made by traditional methods, in which traditional designs are knitted into a seamless garment.

The illustration shows a Chinese pillar carpet which has a pattern continuous around the pillar. It is cylindrical rather than fully three-dimensional but shows the strength of single pattern design. One looks forward to the day when garments can be designed for mass production with such freedom.

As it is fashionable to consider the effect on the world's resources of each new development, it should be remarked that advanced systems do not necessarily use up yet more of these scarce resources. On the contrary in weaving and making up, cloth is cut and wasted. When small-pattern and plain fabrics are made up in garment manufacture the wastage can easily be from twenty to thirty-five per cent, quite apart from the cost of the process. Large-pattern fabrics are still more inconvenient. In knitting there is hardly any waste at all.

Knitting is a more versatile craft than is often realized. For example metal plates can be knitted into a fabric, either as sequins or as titanium plates to form lightweight bullet-proof cloth.

It may seem curious that an advanced technology should be based on so oblique a method as forming a material into a series of long threads and then knitting them together. Why not, you may well ask, form plastics directly into the required

Chinese pillar carpet (19th century)

shapes? Such methods are in fact being pursued in several ways but plastic is not always considered a pleasant material to wear. In the rag trade the point of ultimate choice is the point of sale, and double jersey, for example, is a sympathetic fabric. Knitting will go on for a long time.

What, in outline, might such a system look like? There is space here only to suggest a few of the difficulties and ranges of choice: artificial definiteness hopefully in aid of clarity.

Sketch for a design system

We may picture our designer sitting in front of a screen, rather like a television screen but showing more detail, with several means, carefully chosen for his convenience, for telling the system what he wants it to do: in a computing system all activity ultimately derives from the users. His hand may hold a light pen, or a convenient grip which moves across a sensitive digitizing pad. He will watch on the screen a point of light, or cross, or other convenient marker, which moves across the screen in sympathy with his hand. The stick, which in a pencil joins hand and drawing point, is replaced by a logical or electronic link. He will have a set of buttons for achieving some predetermined results, such as changing colour, and a convenient thumb-wheel for varying magnification of the picture; possibly also a keyboard for typing in information. We shall all have to learn to use keyboards for one reason or another.

The set of buttons is merely a convenient way of telling the system to take one of a predetermined set of actions such as: produce this garment I have just designed; display outlines instead of the shaded picture; proceed to the next stage in the design process; or throw away this design (the wastepaper basket is an important part of any studio). The set of actions made available will vary from one design system to another and can also be made to vary on a given system at different stages of the design process.

One of the most useful features of a computer-aided design system is its ability to recall from memory a library of previous designs, or other information, not necessarily restricted to the manufacturing system in use. A projection system for slides or films can conveniently be combined with the television-type display, the slides being selected automatically by the system on request. The designer, or design studio, may build up a library of standard motifs and reference material.

It may be necessary to restrict access to certain information to designated groups of people. The question of security and control is very important. In a system used by one person this may be a simple matter. If a system is used by many people well defined procedures must be established for adding and deleting from the standard set of designs. There must be no unauthorized alteration of essential information either by accident or on purpose. Certainly it cannot be open to all users of a system to launch production by pressing a button.

Suppose a designer is engaged in designing a dress. Arms, legs and head might be sketched in, indication being made to the system by pressing the right button that such information was for reference purposes only and not a part of the

Opposite:
Shaping things to come

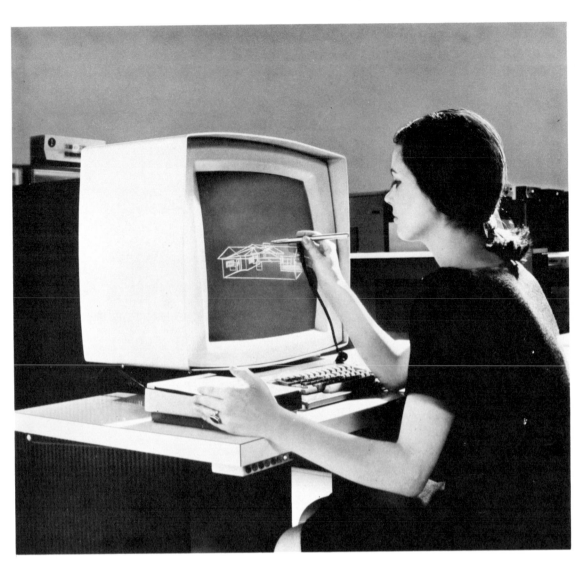

design. The system might be asked to provide from its library a figure, perhaps of a favourite model, who could be put in any position and viewed from any angle, on which the design could be sketched.

A computing system can provide the user, in this case the designer, with a set or menu of choices, and it is much more rapid to choose from an indicated set of possibilities than to remember what can be used. For example a list of the possible types of stitch can be presented on the screen. The designer then makes his choice, defining the stitches or combinations of stitch to be used in different areas, possibly typing in the number of stitches to the inch or other relevant information. The system should ensure the continuity of stitch within the given geometry. The designer can magnify any part to ensure that it is precisely as he wishes, modifying the design until he is satisfied.

The system should ask the designer to make all the necessary choices at each stage, proceeding to the succeeding stage only when it has been provided with all the information which it requires.

The presentation of a menu of choice is a powerful working tool. With a well designed system a teaching manual is largely redundant: the system incorporates a sort of programmed learning as a part of the design system. It may be held that the rhetoric of a subject is not to be learned that way, but it will serve well enough for the three Rs. It would also be of help to a designer transferring from one automated design to another. When the designer is freed from the burden of attention to detail, which grows enormously in a real production situation, his attention is freed for creative design.

The menu of choice is also a convenient method of introducing practical manufacturing decisions. The range of choice of colour can be defined. If a garment is being designed for mass production it may be sensible to restrict the design to the use of materials immediately available and orders can be placed automatically for the manufacture of the yarn needed to meet future requirements.

It is no small problem keeping stockholding and ordering under control in an efficient way. The designer is always trying to widen his range of choice, possibly asking for exotic materials such as sequins or metal plates; but stocks tie up capital, and the wider the range of choice the larger the stocks. It is in the control of such practical matters, together with its ability to shorten the design cycle time and also to produce better designed products, that the economic justification of the computer in design must be found, at least in the low technology industries.

Although design may be done initially for a single garment, in production a range of sizes and fittings will be required. Such a range does not merely require scaling up, but is rather a matter of proceeding according to a prescribed set of rules. Compromise is required if the design is built from a set of stitches which is the same for all sizes of garment. In an automated system the results can be displayed and then adjusted for each size and fitting until satisfactory to the original designer. As a final check a complete set of garments can be produced and tried out.

In principle the model can be moved, by raising the arms and so on, all on the display screen, to see how the garment will adjust, but the conditions might be somewhat arbitrary: perhaps the garment would be considered to be fixed at some points, say hanging from the shoulders. Unfortunately research is a long way from providing the necessary theory.

However one result of closing the design/manufacturing loop, that is to say of providing means whereby a design can be implemented immediately, is that the limitations of displays and of the modelling techniques available at a given time can be circumvented. The picture on the screen may not have the detail or accurately reproduce the colour of the original. The mathematical model of the shape may be far from perfect. But when design has reached the right point the designer simply presses a button to produce an actual garment. Coming from the production line (or equivalent machinery) the garment is typical. The production line is being used as a computer peripheral which is time-shared between production and design.

The most time-consuming procedures in such a system are what they should be: the human problems of creative design and making the essential decisions.

7 Recapitulation

Computers have for some time been used to do the sums in engineering design. More recently the computer's power to handle visual images has been greatly extended to include shaded pictures, colour and powerful theoretical techniques.

With these facilities computer-aided design is being developed in earnest and may be expected to extend its range of use and depth of application. It will gradually become possible to implement an integrated design and production system in one industry after another, starting with relatively simple technologies such as weft knitting.

Design with computers is still expensive. It is made cheaper by time-sharing the facilities between many designers. Even so it is likely to be expensive for some time to come. But the figures look quite different when the design and production processes are considered together as a single entity. The first benefit may well be in shortening the design/production cycle time. Once this is achieved there is likely to be a straight forward cost benefit; although as a result of the radical nature of the changes involved, as was the case with the printed circuit platter facilities discussed in chapter one, it is difficult to prove this in advance.

In order to implement such integrated systems there are many technical problems to be solved in each case; there are also serious problems created by the historical traditions of the division of labour.

In particular the designer can no more rely on the computer specialist to develop design systems appropriate to his needs

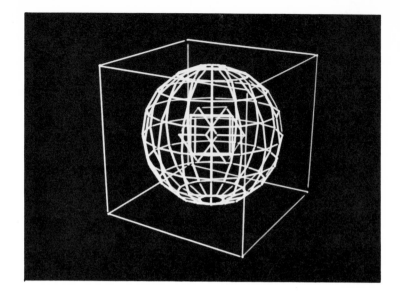

Computer-generated objects: line drawings and shaded pictures.

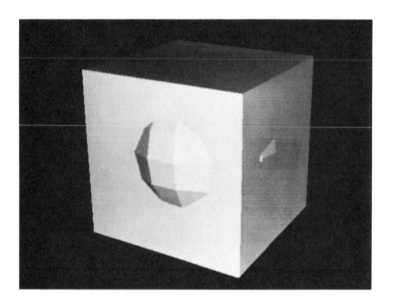

than a wood carver can allow other people to sharpen his tools. The computer specialist does not have the knowledge and cannot acquire it without becoming a designer himself.

With the evolution of such design and production systems the possibility at last exists for mass production to be raised to the level of a craft, well integrated designs being developed by a well co-ordinated team of people, or even by one man: the master craftsman in an age of mass production.

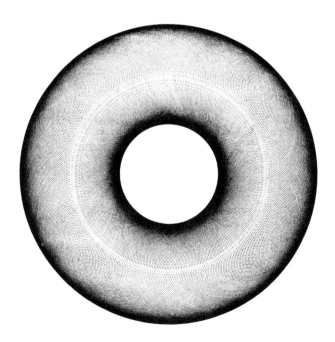

Glossary of terms

Analogue	a smoothly changing variable or indicator such as the hands of a clock as opposed to a digital variable.
Computational geometry	a new type of geometry used in computer aided design of doubly curved surfaces.
Contra-rotating intermittent feed	an advanced knitting system with two circular feed systems rotating in opposite directions.
Coons' patch	the basic element of surface in the original development of computational geometry. It is, or represents, a simple doubly curved surface bounded by four curved edges.
Digitizing pad	a flat rectangular surface on which lines can be drawn and points identified while the digital computer records the action. It can be used in conjunction with a display screen, in which case the hand moves across the digitizing pad while the eye watches the screen on which lines appear corresponding to the lines drawn on the pad.
Display screen	a screen on which the output of a computer can be seen as a pattern of light.
Display tube	the most common display screen used apart from computers is the television set with a cathode ray display tube. Several types of cathode ray display tubes are used in computers.
Flat-bed plotter	an automatic drawing machine which draws on a flat surface.
Hardware	the physical parts which comprise a computer.
Light-pen	a device for identifying a point on a display tube by simply pointing at it. The system has means for measuring where the pen is pointing.
Line printer	a computer peripheral in which a whole line of letters or characters is composed and printed together.
Platter	a complex printed circuit.
Software	the assembly of programs which makes the hardware into a usable system.
Storage tube	a display tube incorporating a physical system for storing the picture automatically at the front of the tube.

Bibliography

Armit, A. P. *Multipatch and Multiobject Design Systems* Proceedings of the Royal Society A.321, London 1971

Buhl, Harold R. *Creative Engineering Design* Iowa State University Press, Ames, Iowa 1960

Critchlow, Keith *Order in Space* Thames and Hudson, London 1969; Viking Press, New York 1970

Doughty, John Carr *Research and Development in Knitwear and Knit Fabrics* Published privately by the Design and Development Centre, Leicester

Fetter, William A. *Computer Graphics in Communications* McGraw-Hill Book Company, New York and Maidenhead 1965

Forrest, A. R. *Curves and Surfaces for Computer Aided Design* Ph.D. Thesis (published privately) Cambridge 1968

Lefkovitz, David *File Structures for On-Line Systems* Spartan Books, New York 1969; The Macmillan Company, London 1969

Pye, David *Nature of Design* Studio Vista, London 1964; Van Nostrand Reinhold Company, New York 1964

Weyl, Hermann *Symmetry* Princeton University Press 1952; Oxford University Press 1952

Wilkes, M. V. *Time Sharing Computer Systems* Macdonald and Company, London 1968; American Elsevier Publishing Company, New York 1968